Union & Confederate

CIVIL WAR HEADGEAR

and related militaria

The Kenneth J. Budny Collection

THE MANUAL OF ARMS
DISTINCTIVE COLLECTIONS SERIES
VOLUME I

by Paul Millikan
and Richard Listenberger

The Manual of Arms Publishing

The Manual of Arms Publishing
317-826-2477 or www.themanualofarms.com

ISBN 0-9765295-0-5

Printed in Canada

A Closer Look

This is not a book devoted to new sources and research. There is no amazing new information offered here. Instead, this is a book that takes a closer look through photography, and a straightforward summary view of information about the pieces in the collection.

This book is the first in an intended series on Distinctive Private Collections in the large general category of Americana. We believe this will be a valuable supplemental reference series because, while public collections are well documented and represented in print, private collections are often overlooked or unknown to the larger collecting fraternity.

While this is essentially a picture book ("a picture being worth a thousand words"), it represents the interests, enthusiasm, research, value, knowledge, opinions, tastes, and dedication of the person who is the collector. All collectors have a bit of the museum director, curator, and exhibitor, in them. Within the realities of one's financial resources, every collector works hard at building a collection. At best, collectors are caretakers of their collections for a few score years, and then at some point, the collection is sold and broken up to become parts of many other collections. An auction catalog is often the only source of public recognition a collector realizes in a lifetime.

We judge a collection to be important because it speaks to a lifelong love affair with the things that comprise the collection – important also because some of the pieces in the collection are unique and outstanding for their condition, history, or details. Such pieces are best presented in the environment of the collection that is their home. The provenance and details of each such piece needs to be established and preserved for the benefit of future collections. This book is an effort to do just that.

The Kenneth J. Budny Collection

Ken Budny has been a serious collector for over 50 years. During the period from 1958 through the early 1980's, he built one of the largest and finest collections of Ethan Allen firearms ever assembled. He won many awards for displaying his collection at various shows over the years. Life circumstances caused him to sell that collection in the late 80's, and for a few years he was unable to settle on one specific new collecting agenda.

In the early 90's, Budny began to focus on building a new collection in 19th century American Artillery. This was a broad collecting interest, and resulted in the acquisition of all sorts of artillery-related items, including headgear.

Civil War headgear and uniform cloth emerged as Budny's collecting passion . . . "especially headgear. The hats seemed so personal. Often the vet continued to wear his wartime hat when attending reunions or even out in the field . . . so personal" said Ken, when speaking of what brought about this emphasis in his collecting.

His plans were to share this new collection by displaying at shows beginning in 2004, but poor health prevented fulfillment. Friends did display part of his collection at the fall Gettysburg Show in October 2004. They also displayed for him at the November, 2004 24th Annual Capital of the Confederacy Show in Richmond, Virginia. The Gettysburg show did not give awards. However, the collection was very well received by the collecting fraternity. At Richmond the display did win the coveted M. E. Mason, Jr. Best of Show Award.

Ken Budny is well known and much respected by collectors and dealers around the U.S. He is a member and former officer of the Michigan Antique Arms Collectors; a Life Member of the Ohio Gun Collectors Association; a member of the Pennsylvania Antique Arms Collectors; and a former member of the Antique Arms Collectors. His winning smile and personality have been a staple of shows around the country for many years, and it is hoped to be for many more to come.

One of the chief characteristics, and perhaps the most outstanding thing about this collection is its great variety. When the collection was displayed, it was under the title "Not as Simple as Blue and Gray – Civil War Headgear." The collection is very colorful and reflective of what we see in photo collections such as Miller's Photographic History of the Civil War. These photos reveal volunteer regiments, companies, leaders, and individuals wearing a great variety of headgear, even and almost always within the same group.

The pictures of newly formed and mustered in regiments show them lined up and dressed to the 9's in near regulation headgear. However, pictures taken later reveal that regulation has given way to personal taste and comfort. After a year or two, at least 20% are wearing non-regulation headgear.

Variety makes this a fun collection: fun to look at; fun to research in period photos; fun to compare, and just plain fun. So enjoy as you survey and read these pages.

Richard Listenberger and Paul Millikan
Authors and Publishers

Photo Courtesy of Richard Caldarelli

2004 24th Annual Capital of the Confederacy Show in Richmond, Virginia

M. E. Mason, Jr. Best of Show Award

1

Model 1864
US Light Artillery Shako

Officially known as the Model 1864 US Light Artillery cap, it has come to be referred to by collectors as a Light Artillery shako. This was for enlisted men and is of dark blue wool over a hard pasteboard body with a black leather crown.

Figure 1, shows all of the original insignia, red horsehair plume, cords and hangings. The left side view (Fig. 2) shows the distinctive slight curve of the back. This is much less severe than the curve on a Model 1851 shako.

The visor is of unbound heavy leather and has a unique shape (shown clearly in Fig. 3). It bulges out on each side to be actually wider than the body of the hat. The complete sweatband is 2 1/8" wide.

The close-up of the label inside the crown reads: “W. C. Dare, Manufacture, 21 North 2nd Street Philadelphia” (Fig. 4).

This is definitely Civil War as Dare had the only contract to make these shakos in 1864 and 1865. After that, all contracts were with Horstmann Bros. & Co. of Philadelphia beginning in 1867.

2

4

3

5

"Ringgold Cap" or Model 1858 US Light Artillery Dress Cap

This shako for officers has some confusing nomenclature. The nickname for it was the "Ringgold Cap" and it is a reference back to the shakos worn before the Mexican War and up to 1851 by Maj. Samuel Ringgold's company of artillery. Also referred to as the Model 1858, because in 1858, when all officers of all branches were ordered to wear the officer's Hardee slouch hat, many artillery officers preferred, and continued with the older Ringgold cap. Early in the Civil War it became clear that these were impractical in the field and their use was virtually abandoned until 1864, when the new Light Artillery cap for enlisted men came into use. This brought a revived use of the Officer's Dress "Ringgold Cap" once again.

The front view (Fig. 5) shows the original bullion embroidered insignia and red horsehair plume in a distinctive brass holder. The heavy braided cord is gold bullion.

6

The body of the shako is of hard pasteboard and covered with black beaver on the outside. The crown is of polished black leather.

Figure 6 shows the more drastic curve of the back (similar to that of the 1851 Shako), and the Eagle "A" buttons holding the chinstrap and the braid.

The inside view (Fig. 7) reveals the red and black polished cotton striped lining, now faded to a dull pink and black stripe. The underside of the bound, heavy leather visor is green. The small 3/4" x 1" glued-in paper label has written in pencil 22 1/2". The leather sweatband is 1 7/8" wide.

7

9

American Militia Shako for Artillery circa - 1840's

Militia hats of this style were used by local volunteers in many locales both north and south. When the Civil War broke out, many such units enlisted en masse and wore their militia uniforms off to the larger army. Obviously, this didn't last long as they were highly impractical and were originally more for dress and local parades than anything else; such units were also mostly three-month enlistments. When in this top hat style, they were called shakos.

Figure 9 shows the brass eagle insignia universally used from the 1830's through the Civil War on various kinds and styles of headgear. The hat is of firm heavy wool felt and trimmed with red wool piping.

The left side view (Fig. 10) features a jaunty red, white, and blue cockade with an eagle "A" button in the center. The two tassels are of gold bullion and behind them is a gold bullion and sequins 5-point star attached to the red hatband.

On the inside (Fig.11) it has no sweatband. There is no stitching line or shadow to indicate that there ever was a sweatband.

The close-up of the label reads "A. P. Megrath" over a mounted soldier with sword, and then additional information. "Keeps constantly on hand an assortment of MILITARY WORK consisting of caps, belts, holsters, valises, housings, breast girt, etc., etc. , also saddles, bridals, harnesses, trunks, whips, etc. etc. Greenfield, Mass." (Fig. 12).

The printer is "Herald Office, Greenfield".

The label makes this a very outstanding example of the militia top hat/shako. Most of these are without a label, much less the label of a military maker/supplier.

10

A. P. MEGRATH,
KEEPS CONSTANTLY ON HAND,
MILITARY WORK,
CAPS, BELTS, HOLSTERS, VA-
LICES, HOUSINGS, BREAST-
GIRTS, &c. &c.
ALSO, SADDLES, BRIDLES, HAR-
NESSES, TRUNKS, WHIPS, &c. &c.
Greenfield, Mass.
NO.

12

11

13

Model 1858 Officer's Hat (Officer's Hardee)

Adopted in 1858 this style was worn by many union officers throughout the Civil War. This example conforms to regulations in every respect except one. Regulations called for the ostrich plume to be on the left side, and this one is on the right. During the early part of the war, mounted officers learned that the plume stayed on the hat better when protected by being tucked behind the folded right side. Examples are found with the ostrich plume on either side. This example is in near mint condition.

Figure 13 reveals an embroidered artillery insignia with the number "2" in silver in the center. It was worn by an officer of the 2nd artillery. The gold bullion hat cord is long and heavy. The hat is of heavy black felt and is fairly stiff.

The right side view (Fig. 14) has a really spectacular embroidered eagle insignia. Both of the insignia are embroidered on black velvet, which is mounted on a tin oval with two attaching loops on the body.

The inside has a complete linen lining and an oilcloth sweatband, which is 2" wide (Fig. 15).

Provenance: Several years ago (late 80's) this was purchased from a Wisconsin widow. Her husband had purchased it for his collection in the 40's, and he treasured it because it was in near perfect condition. He kept it in a hatbox so it never even got dusty. From her, it went into the prestigious Robert Walter Collection in Chicago. When it was sold again it went to Mr. Budny.

14

15

16

A Child's US Infantry Slouch Hat

This is typical of the many that were purchased by Federal Officers and their wives for their small male children. Often, they were outfitted in complete small uniforms. So junior was just like Dad.

In Figure 16, it has no front insignia, but the hat cord is knotted several times to fit the small head size. The hat is made of pressed black felt and is the weight and consistency of a regulation enlisted Hardee. The brim is bound with black velvet which has dye faded to very dark brown.

The right side view (Fig. 17) shows a small black ostrich plume and a brass eagle side plate like those on enlisted Hardee hats.

The inside view reveals that it has no lining (and never did). The complete sweatband is of brown leather and 1" wide (Fig. 18).

17

18

19

Confederate Georgia Kepi

Made of a dark gray wool and bound around the base with dark brown jeans cloth, this has typical confederate characteristics.

From the front view (Fig. 19) it has a heavy tarred chinstrap, and the visor is of tarred oilcloth over pasteboard with a bound edge of the same material.

The left side shows Georgia side buttons attaching the chinstrap (Fig. 20).

The inside view (Fig. 21) reveals no sweatband and it likely never had one. There is quite a lot of sweat and dirt staining where the sweatband would have been. It was worn a lot with no sweatband. The lining is of natural unbleached cotton and all hand sewn.

The close-up of the identification marking reads: "J. J. Cox, Co. H 42nd Ga." (Fig. 22)

This kepi belonged to 1st Lt. John J. Cox, Co. H, 42nd Georgia. He was taken prisoner in July 1863, at Vicksburg, Miss., when the garrison there surrendered to General Grant. He was later paroled and returned to duty only to be severely wounded in the leg at Missionary Ridge in 1864. His wounds caused him enough disability that he was finally forced to resign and leave the service in 1865.

20

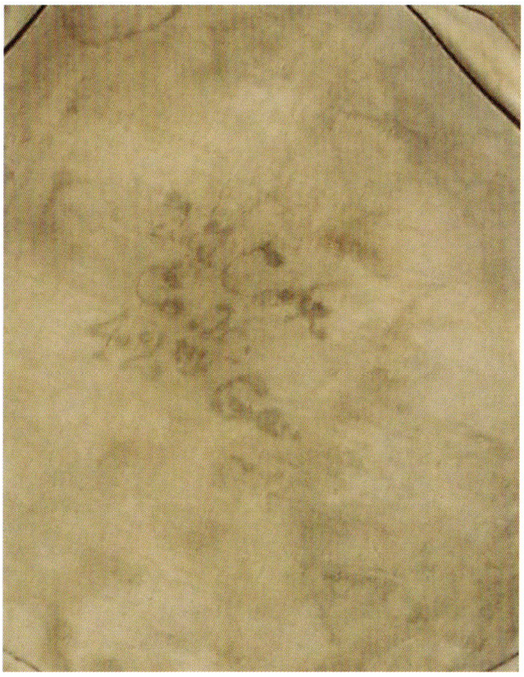

22

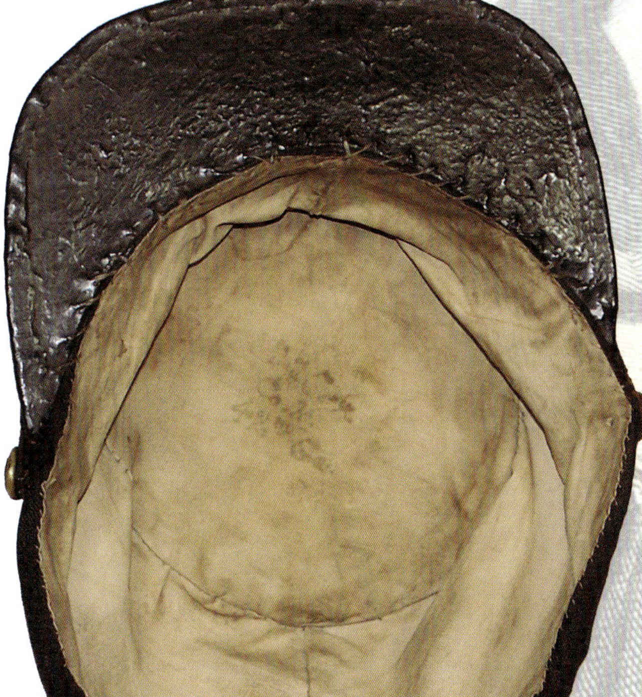

21

23

Confederate (Regulation) Enlisted Gunner's Kepi

Made of Red Wool with a blue wool band around the base, this is exactly what C.S. regulations called for in artillerymen's caps.

The front view shows a wide black oilcloth chinstrap and a bound, tarred visor. The kepi measures 4" high in the front (Fig. 23).

Figure 24 shows plain US eagle side buttons, which are often found on C.S. kepis.

The inside has a green underside on the visor. The cotton lining is only on the sides. The crown is exposed pasteboard and was never lined. The sweatband is 1" wide and made of black oilcloth (Fig. 25).

24

25

26

Confederate South Carolina Officer's Kepi

With a gray crown, red body, and gray band around the base, this is exactly like several known examples in reference publications and museum collections.

The front (Fig. 26) has several distinguishing features. It has an embroidered Palmetto Tree insignia on the front. This is embroidered on red wool rather than the more typical black. The chinstrap is of black oilcloth. The visor is bound with and covered with black oilcloth as well.

On the left side (Fig. 27) it has flat, one-piece, floral buttons holding the chinstrap.

In Figure 28 we find black oilcloth inside the crown, a brown oilcloth sweatband, and the side lining is brown polished cotton with a very faint narrow stripe woven in. The sweatband is 1" wide as is the gray wool band around the outside base.

27

28

29

Confederate Virginia Veteran's Slouch Hat – circa 1880-1900

This is not U.C.V., but rather Confederate Veterans. This is the same group that published the Confederate Veteran Magazine.

The front features a gold washed wreath insignia with the silver initials C.V. in the center. A gold bullion hat cord is around the base and there is a clear 1 1/8" wide shadow where the hat originally had a hatband. This was very likely made of silk, and separated and deteriorated as silk will (Fig. 29).

The right side (Fig. 30) is highlighted by a striking, three tiered black cockade with a large pewter Virginia button in the center.

The inside has a complete 1 5/8" wide sweatband. This has a makers' mark printed in gold, but it is too worn to be made out. The hat never had a lining (Fig. 31).

30

31

32

Confederate Louisiana Slouch Hat

This was likely an officer's slouch hat, because of the outstanding cockade on it. It has a very wide (4 1/8") brim. The edge of the brim is not perfectly round and smooth, but is instead irregular, and was trimmed by hand. It appears that the edge was bound as was typical, and the binding became worn out, and the edge was trimmed just inside the stitching line. There are also 2 small places on the right edge, which could be the evidence of edge damage, and that was the reason to trim the whole edge by hand. The trimming may well have been done after the war.

The hat is made of heavy, tan, pressed wool felt, and is relatively stiff.

In Figure 32, we see a crimson wool hat cord. This is a very dark red, similar to an officers' sash color. This is federal and for enlisted men.

The left side (Fig. 33) has a great Louisiana Pelican Cockade (see the close-up Fig. 35). The center is red and the pelican is pewter, while the cockade ribbon is three-tiered and black. This is an extremely rare and pre-civil war insignia. There is also a faint 5/8" wide shadow left by a hatband, now missing.

The inside has no sweatband, but there remains a 1 5/8" wide shadow impression from where a sweatband was. This hat was never lined (Fig. 34).

33

35

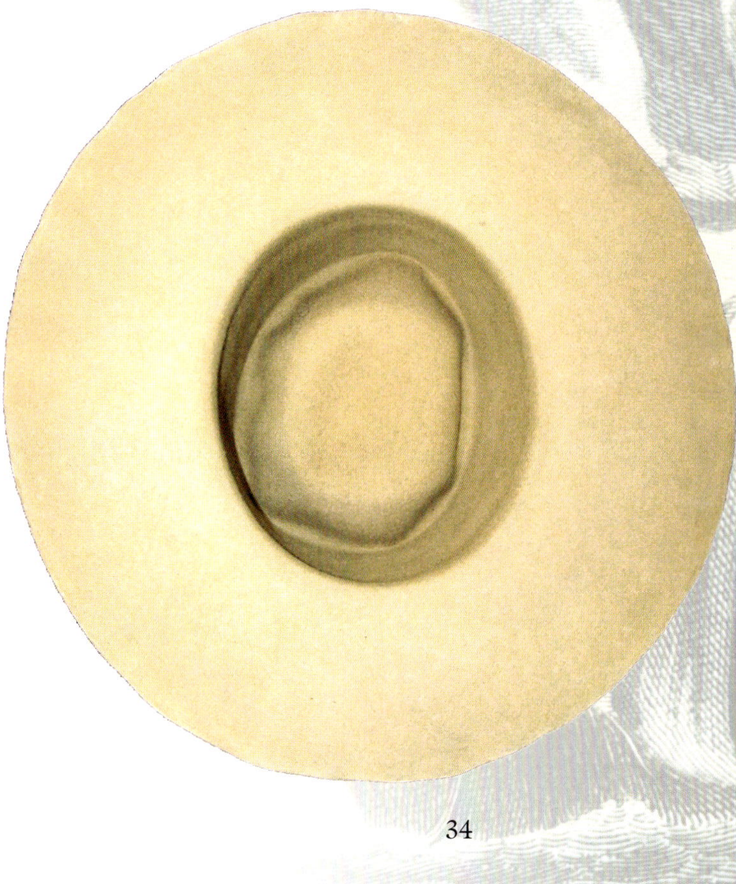

34

36

US Artillery Officer's Camp Hat – circa 1861-1865

This hat is of a known Civil War pattern, but is quite rare and very difficult to describe (even with pictures).

The first picture (Fig. 36), shows the right side with insignia showing, and the hat folded flat and forming a half circle. The insignia are for the 5th Artillery, and are stamped brass in false embroidery.

When opened and worn, it appears from the front, as seen in Figure 37. Opened, it becomes a slouch type hat, but the wide brim is only on the sides. There is no brim on the front and back. Also, when opened, the insignia are not seen clearly as they are now on the underside of the brim.

A side view is seen in Figure 38. Again, this is strange looking, because there is no brim front and back.

Figure 39 reveals that the inside of the hat has a complete brown, polished cotton lining. The sweatband is black leather and 2" wide.

A similar example of this style may be seen among the camp or smoking caps pictured in *Echoes of Glory*, page 188.

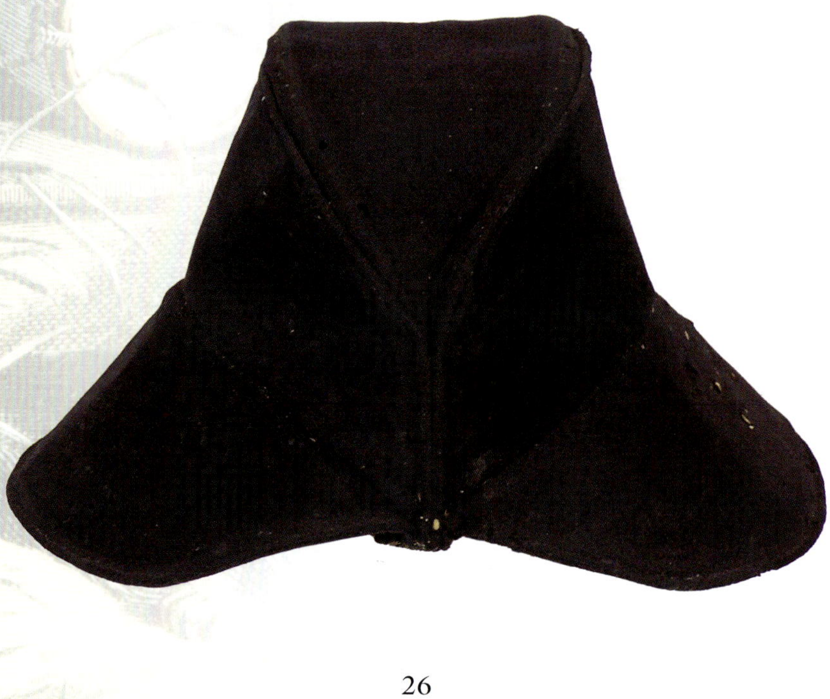

37

38

39

40

Confederate Infantry Officer's Slouch Hat

This is clearly non-regulation in almost every respect, but again, that is not unusual on either side during the Civil War.

Confederate regulations did not call for insignia on the front of hats, but many, particularly those who had served in the "Old Army" before the war, wore them anyway. This embroidered Infantry Horn insignia (Fig. 40) is circa 1858. The officer's hat cord is extremely long, so that the acorns reach over the edge of the 3 5/8" wide brim.

The side view (Fig. 41) emphasizes the length of the cord and shows a hatband that is 1" wide. The hat is made of heavy, and now stiff with age, light brown wool felt, and has a bound brim. Originally, the hat would have been quite soft and pliable.

Inside the hat (Fig. 42), we find a 1 7/8" wide, brown leather sweatband, which is well worn, but complete. These is no evidence this hat ever had a lining.

41

42

43

New York Artilleryman's Kepi – circa 1890

This kepi is typical in construction of those from the late Indian Wars. It is unique for its bright red, wool crown. Made for artillery officers, the body of the hat is of dark blue wool.

The front of this kepi (Fig. 43) has a stamped brass crossed cannon insignia and a bullion cord chinstrap.

In Figure 44, we note the state of New York side buttons. The visor is of very thick, unbound, black leather.

Figure 45 shows the inside of the kepi, which has a complete lining. The crown is lined with black polished cotton and it bears the illegible remains of a printed label. The sides are lined with brown polished cotton. The sweatband is complete, made of black patent leather, and bears a size "7" label.

44

45

46

Federal Artilleryman's Civil War Forage Cap

This fine quality hat has identification to a New York Artilleryman.
Figure 46 shows the hat extended to its full 4" front height. It has false embroidered brass cannon insignia on the front, and a false embroidered number "9" toward the front on the crown.

The side buttons (Fig. 47) are US eagle "A"s, and the chinstrap is 9/16" wide black leather. The edge of the black leather visor is bound.

The underside/inside view (Fig. 48) shows a scarlet silk lining on the sides, and a black linen lining in the crown. The complete sweatband is 1 5/8" wide and made of polished, brown, Moroccan leather. The underside of the visor is green.

The identification (Fig. 49) is in ink on a small piece of paper glued in the crown. It reads "Elizer Benson, 9th Regt. NY Arty". He enlisted late in the war at age 34 in Rochester, NY. His service was from Sept. 22, 1864 through June 31, 1865. He saw service and action at Cedar Creek, Petersburg, and Sailor's Creek, all in Virginia.

His tintype Civil War photo (Fig. 50) accompanies the hat.

47

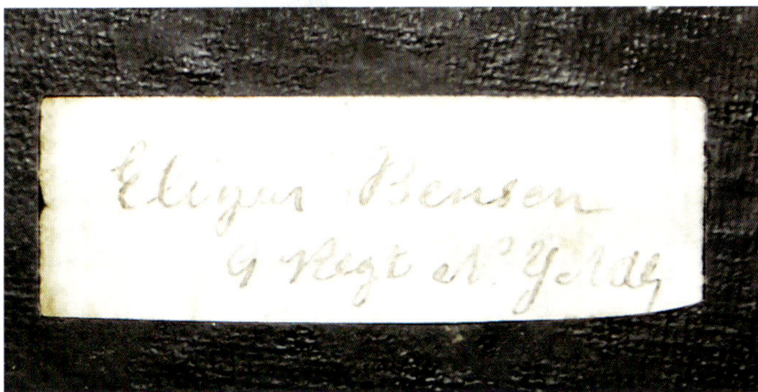

49

48

50

51

A Large Indiana Grouping: Three Generations of Artillery Officers in the Thompson Family

There was much too much in this group to try to picture everything. Figure 52 shows some of the more important items from the group.

The first member of the family to be in the artillery was Doctor James Livingston Thompson (Fig. 51). He was Civil War Surgeon of the 4th US Heavy Artillery (colored), and for a period of the war he was Medical Director for the army for Western Kentucky. A cdv of him with his medical staff sword and standing in uniform, is in the family album, which is part of the group (Fig. 52). The album also contains another cdv of Thompson in uniform (a portrait). The officer pictured in the album (Fig. 52) is his brother-in-law, A. H. Tevis, 2nd Lt. 37th Indiana. There are several other cdv's of Tevis in the album, but no others are in uniform.

The group also includes Thompson's casebook, with his handwritten entries concerning 9 different soldiers that he treated in 1863 (some with amputations).

His specialty became eye surgery and the 2 casebooks in the group contain several of his drawings of the eye in color, and detailing procedures. He is quite famous in Indiana, as he performed the first cataract surgery in the state, and one of the first in the nation. He became a professor of diseases of the eye at the Medical College of Indiana.

Finally, the group included a dozen steel engraving portraits of him, circa 1880; his schedule, date/notebooks for 1902, 1904, 1909, 1910; his framed, printed-on-silk "Fellowship Pledge of the American College of Surgeons;" a wallet inscribed to his partner, Dr. John H. Oliver; and two copies of the memorial book to Dr. James Livingston Thompson 1832-1903, and his son, Dr. Daniel A. Thompson 1862-1904.

Dr. Daniel A. Thompson is the second-generation artillery officer. He was an Indiana Wars Surgeon and Spanish American War Surgeon for the first Brigade, Indiana Light Artillery. His Indian Wars kepi (Fig. 52) with embroidered insignia First Artillery I.L.A. (Indiana Light Infantry) is in the group. There are also three pictures in his part of the group:

1. A cabinet card of Major Daniel A. Thompson in frock coat and shoulder knots
2. A cabinet card of him standing in full dress uniform with his sword and 1881 dress helmet, and
3. An albumen photo of him with 5 of his staff officers (all in uniform) from the Spanish American War.

He followed in his father's footsteps becoming an eye surgeon and Professor of Diseases of the Eye at the Medical College of Indiana. He died 9 years before his father.

The third generation artillery officer was grandson to the first, and named for him. From this second James L. Thompson, the group includes: only his commission, a First Lt. of Field Artillery Officers Reserve Corps, dated Nov. 27, 1917.

This group contains a lot of material, records, and artifacts from a most remarkable Indiana family.

52

53

Identified Artillery Officer's Kepi – circa 1890

This very striking Indian Wars kepi is marked on the sweatband "E. S. Curtis". He was Edwin S. Curtis and he was a Private in Co. D, 48th New York Volunteer Infantry from Sept. 1, 1861 to Sept. 16, 1863, when he was discharged. His discharge from the volunteer army coincided with his entering the US Military Academy at West Point on Sept. 16, 1863.

Curtis graduated from the Point on June 17, 1867, and became a 2nd Lt. assigned to the 2nd US Regiment of Artillery. In the regular army, he was promoted to 1st Lt., on Feb. 10, 1873, and continued service with the 2nd Artillery. He became a Captain of the 2nd Artillery on Aug. 27, 1896. His final promotion came on May 8, 1901, to the rank of Major, US Coastal Artillery. He died Nov. 4, 1901, still in the army.

This is an outstanding example of a regular army officer's kepi, circa 1890. A private purchase, as all officers uniform items were, it is unique for its scarlet color. Officers must have been granted some latitude in their headgear choices, as this is certainly not regulation.

The red wool body has a black wool band/boarder around the base (1 1/8"). There is a single strand of gold bullion lace in a quatrefoil pattern on the crown (Fig. 53), plus edging on the black band and vertical lines on each side and in the back. This single strand indicates the rank of 1st Lt.

Our attention is also drawn to the embroidered artillery insignia. This is a Civil War period insignia of gold bullion embroidery on black velvet and all mounted on an oval of tin, with 2 attaching loops on the back. The hat maker would have ordered this from a dealer in US Military surplus, or Curtis may have acquired it from such a source himself. At any rate, the insignia is much earlier than the kepi.

The hat has US Staff Buttons on each side attaching the leather chinstrap (Fig. 54). The visor is heavy black leather and is unbound.

Figure 55 reveals the inside, which has a complete polished cotton lining. The sweatband is 1 3/8" wide brown leather. The underside of the visor is dark green.

54

55

56

"Pork Pie" Style (Non-Regulation) US Artillery Officer's Slouch Hat – circa 1861-1865

This is one of the best known and readily recognizable styles of Civil War non-regulation headgear, always associated with General Phil Sheridan, who wore a hat of this style in his 1864 Valley Campaign in Virginia. The hat is made of black beaver fur felt over cardboard and linen.

The front of the hat (Fig. 56) is dominated visually by a bright bullion cord with large acorns and the false embroidery brass cannon insignia.

Figure 57 reveals that the brim is bound and the hatband is a very wide two inches, which is half the height of the crown.

On the inside the lining is missing from the crown, but the black side lining is complete. The most interesting thing is that the sweatband is a two inch wide quilted portion of the lining. The quilting is in a large diamond pattern (Fig. 58).

57

58

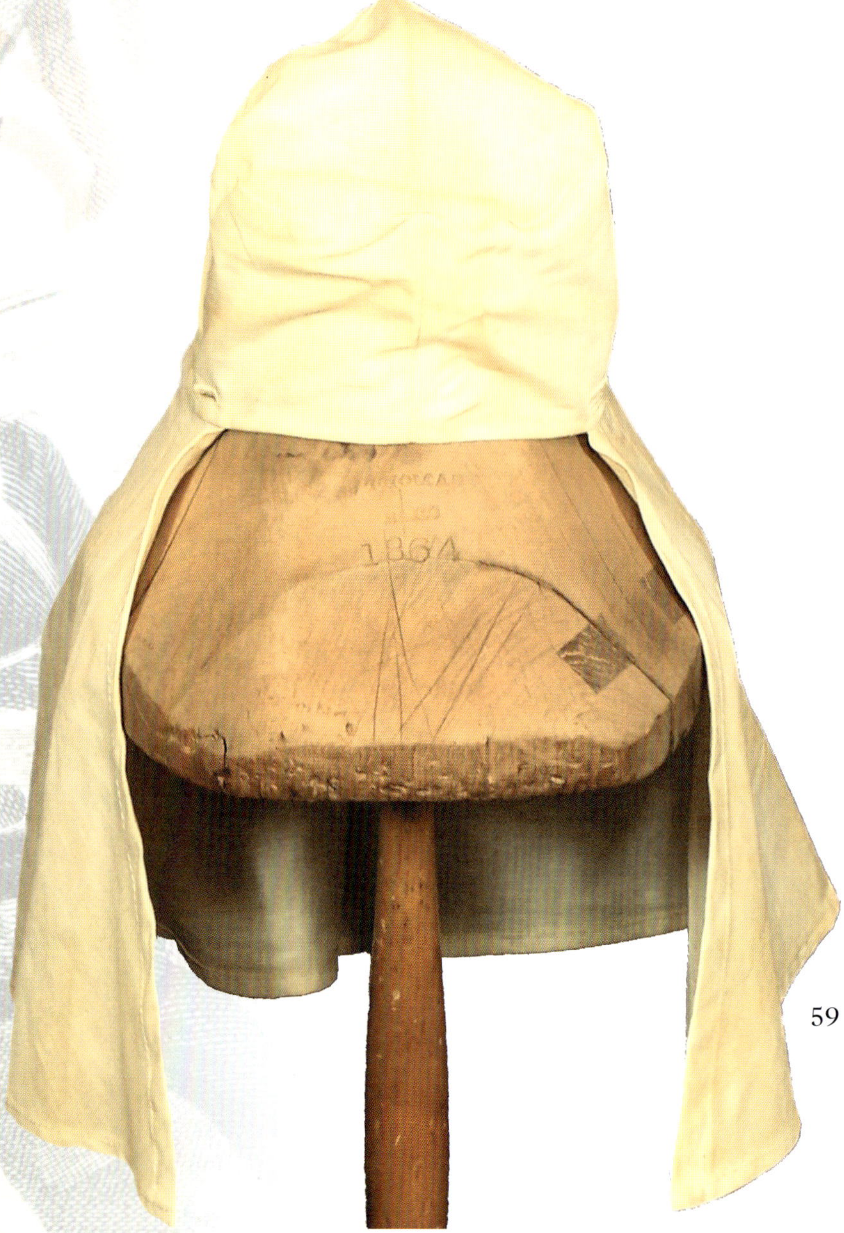

59

Civil War Federal Havelock

These were for both enlisted men and officers, but were never very popular with troops on either side. Essentially, these were to be a shield from the sun. The top portion was formed to fit over a forage cap or kepi. It has two buttonholes on each side, so it can be attached to the cap and not easily blow off. Two buttonholes on each side are to allow it to fit on a variety of cap sizes. The idea was a universal or one size fits all Havelock.

This one is in un-issued condition and made of a soft unbleached linen. In Fig. 59 it is viewed from the front on a hat form used in making Havelocks. Fig. 60 shows it again on the form and from the left side, while Fig. 61 is taken with it on a hat stand and hanging straight down, also from the left side. Figure 62 is of it on a hat stand and hanging straight down.

Provenance: A verbal background attributes it to having been sold as surplus by Bannerman's in the 1930's-40's, and since it is in un-issued condition, this is reasonable.

It was in the prestigious Robert Walter Collection for several years before becoming part of the Kenneth Budny Collection.

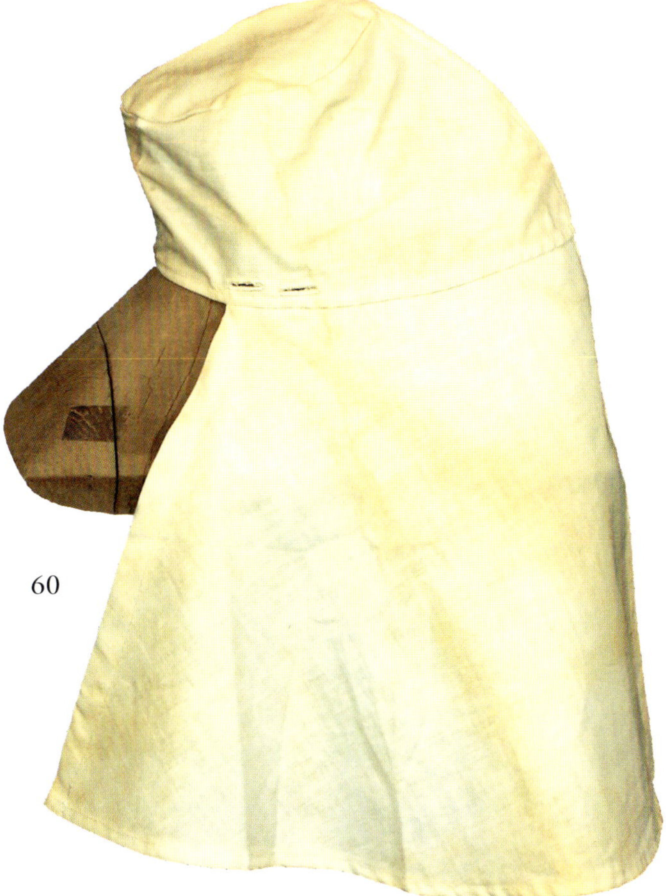

60

This rare Havelock wooden form is made of pine and very well marked. "J. THOMAS – BALT. 1864".

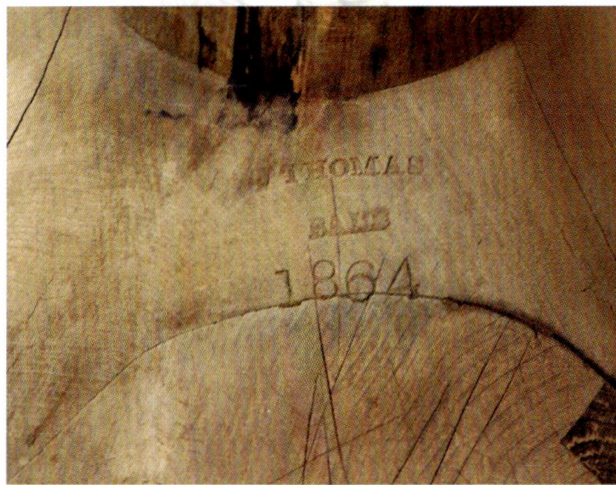

61

62

63

64

Drummer Boy's Uniform Coatee & Forage Cap – circa 1858-1865

Made for use with the Artillery, both pieces are piped and trimmed in red. The tail coat still has its original epaulets. The coat and hat are of a fine woolen fabric, which is green in color. On some spots on the inside and at some stitching edges, the fabric is darker, a virtual green/black. We believe the fabric was originally darker and either faded in the light or, more likely, the dye was unstable and faded with time.

Collectors like the same green in the lining of C.W. officer's frock coats, and don't like it much when the lining is black. Stop and realize that the combination of some woolen fabrics and black dyes were unstable and over time faded to green. Most linings in officer's frock coats of the Civil War were black originally.

Figures 63 and 64 show the coat and cap together from the front and side. Let us now consider the pieces separately.

65

66

The militia coat, referred to as either a tailcoat or a coatee is piped on the front with red worsted lace (Fig. 65) in parallel horizontal lines on the coat bib, as was required in the regular army for musicians. The piping is around the top edge and base of the collar as well. The sleeve cuffs are decorated with matching red fabric. There are three US eagle buttons on each cuff and there are nine large US eagle buttons down the front.

On the back (Fig. 66) we see that the coat tails are decorated with the same red fabric. There are 8 large US eagle buttons decorating the tails.

Pictured lying down in Fig. 67, we can view the inside of the coat. The upper, body portion of the coat and sleeves are lined with an off-white cotton. The chest lining is quilted in a large striped pattern. There is an inside breast pocket on the left. It has no stampings or markings in the lining. The tails are lined with black cotton.

67

68

The forage cap is made of the same colors and fabrics as the coatee. You will note, in Fig. 68, that there is a 1 3/8" wide red band around the base and it has a bound, thick black leather visor. In Fig. 69, we discover that the vertical lines of red piping are double; that the chinstrap is black leather, and the side buttons are US eagles. The height of the front of the hat is a full 4".

The red crown (Fig. 70) is boldly piped with an outer large circle and an inner 4-loop design, done in wool cord that is also green.

The full lining of the cap is the same color cotton as in the body of the coat. It is complete, and the crown is quilted in a large diamond pattern. The brown leather sweatband is 1 3/8" wide and complete (Fig. 71).

69

70

71

72

The epaulets, shown in Figure 72, are original to the coat and made of the same red fabric as used on the coat and hat. The red fringe is interspersed with silver bullion strands. The buttons are US eagle "A"s. The undersides are of polished cotton and each has as long brass wire hook for attaching to loops on the coat shoulders.

74

1885 U.C.V. Reunion Flag

This St. Andrews Cross or Battle Flag is printed on thin cotton. Stenciled very faintly on it in black is "U.C.V." and the date "1885". It is framed. The frame measures 22" x 18" (Fig. 74).

75

C.S.A. St. Andrews Cross Flag – circa 1880's

This small (14" x 10 1/2") printed-on-cotton flag is framed (18 3/8" x 14 3/8"). This was likely made for reunion parades and celebrations (Fig. 75).

76

C.S. Veterans Commemorative Flag

This is all hand made. The red cotton stripes are hand sewn on the white cotton rectangle of the flag and the white cotton stars are hand sewn onto the crossed stripes. Painted on the white triangles in black is: "ALA", "1ST", "VOL", "CAV". The flag measures 30" x 22" and has all three original pairs of ties for attaching to a pole (Fig. 76).

It is highly personalized as well. The back has a 12" diameter circle hand sewn in the center (Fig. 77). Embroidered on this are 17 names of the men and women who made the flag for the post. Four of the names have Ohio locations embroidered as well. Two were Cleveland, Ohio, and the other 2 nearby. There were many U.C.V. Posts in the north, and the one at Cleveland was sizable. The largest U.C.V. Post in the nation was at Chicago. Reconstruction was apparently a time to move north.

77

78

Florida U.C.V. Slouch Hat & State Flag – circa 1900-20

These two pieces came together out of the same Florida Estate.

The beige slouch hat has U.C.V. in a wreath as its insignia at the front (Fig. 78). Another insignia slightly to its left is of two palm trees, and is stamped from sheet copper. This is a very unusual and scarce Florida insignia. The hat has a post C.W. Officers hat cord. There is a shadow from a 1" hatband, now missing. The brim is bound.

The hat is pictured on top of a folded State of Florida flag. The stencil marking on the grommet edge is "U.C.V. Camp 10 Pensacola, Fla."

The flag is printed and in full color as depicted in Figure 79, which is a close-up of the state seal at the center.

Printed on the inside of the hat (Fig. 80) is the name "W. A. Rowls". He was an officer of the U.C.V. Post #10. The sweatband is gone, but a distinct shadow indicates it was 1 1/2" wide.

79

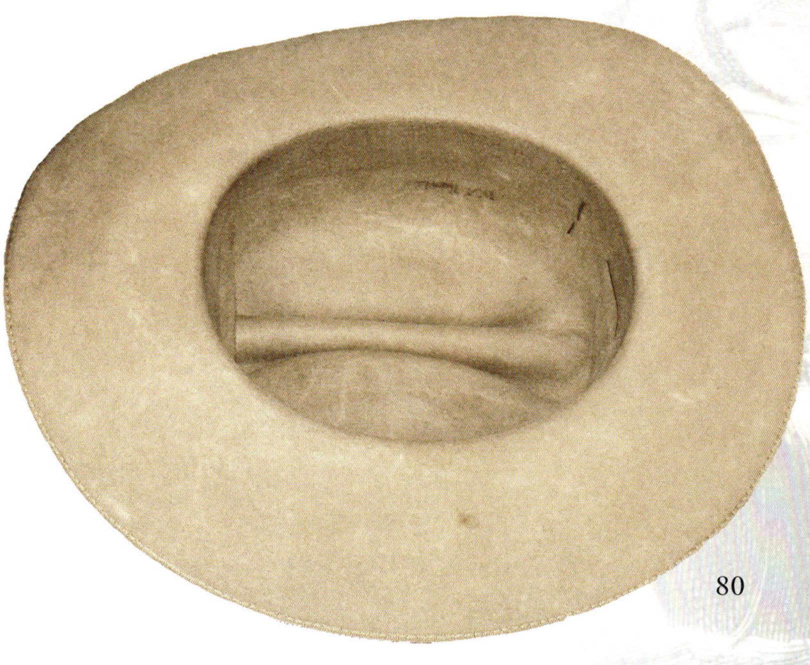

80

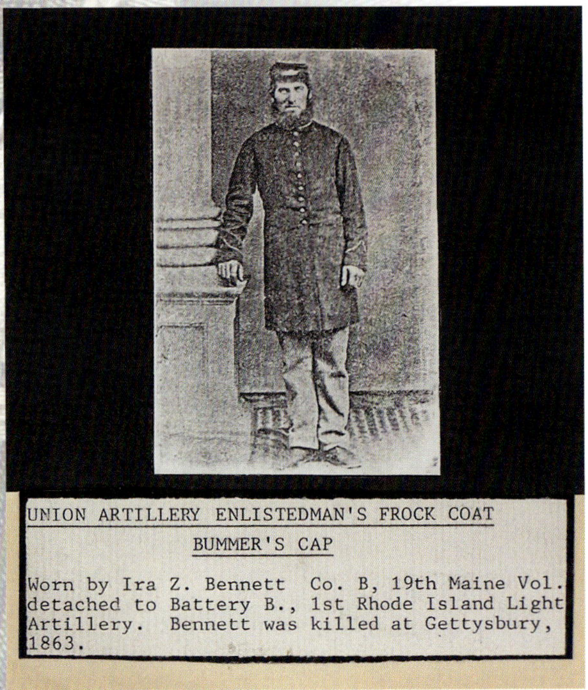

82

81

Identified Rhode Island Enlisted Man's Artillery Frock and Bummers Cap – circa 1861-63

Pictured in Fig. 81, is the enlisted artillery frock and forage cap of Pv't. Ira Z. Bennett, Co. B, 19th Maine Volunteers detached to Battery B, 1st Rhode Island Light Artillery. Bennett was killed at Gettysburg in 1863. A photocopy picture (Fig. 82) of Bennett in his Maine Infantry frock and forage cap accompanies this coat and hat.

Provenance: This identified grouping was offered in the 1970's by The Gettysburg Suttler, George Lower, and it was purchased then by Robert Walter and became part of his collection. When The Horse Soldier sold the Walter Collection, this outfit was purchased by Mr. Budny.

Robert Walter received the photocopy picture from Mr. Lower at the time of his purchase. Any information regarding the location of the original photo has been lost.

Let us proceed to consider the frock and cap separately.

83

84

85

Ira Z. Bennett's Rhode Island Enlisted Artillery Frock

Figure 83 gives us a full frontal view of the coat with the cap sitting on the collar and shoulders. The frock is a full-length (unaltered) 9-button style. The buttons are US eagles. It is a state issue coat from Rhode Island and not Federal issue.

From the left side (Fig. 84) we see the high standup collar and sleeve cuffs all piped in red wool.

A look at the back (Fig. 85) reveals the same cut and tailoring as a regulation coat. There are 2 US eagle buttons on the left cuff, but one is missing from the right cuff.

Looking at the inside (Fig. 86), the chest is lined and padded with a faded to green cotton lining. There is no inside chest pocket. The original sleeve lining is a checkered cotton fabric.

Each tail has a deep pocket, accessible from the outside. The pocket material is brown polished cotton like that seen in many Civil War cap linings.

The entire coat is hand sewn. If there was originally a full lining, all evidence of it has eluded our inspection.

86

87

Ira Z. Bennett's Enlisted Artillery Forage Cap

This Bummer's Cap (Fig. 87) is 4" tall in the front. It has a wide leather chinstrap. The only insignia on it is large crossed cannons of stamped brass on the front of the crown.

Figure 88 views the cap from the left side and it has US eagle side buttons. The visor is thin black leather and unbound.

The interior has a full brown polished cotton lining. About one fifth of the brown leather sweatband is missing, and the rest is very worn.

88

89

90

7th Virginia Cavalry Officer's Slouch Hat – circa 1858-1862

This hat is constructed of black beaver felt over unbleached linen, over stiff hat board. This looks a lot like a Hardee hat, but it's made much differently, and the board material is so stiff that it would break if the side were turned up.

The crown is 6" high and the brim is 4" wide.

In Figure 90, the Virginia designation is clear with the 3/4" high, lead filled, brass letters "VA". The crossed sabers insignia has turned aqua colored with corrosion. The insignia is the 1855 pattern of false embroidery that was used into and through most of the Civil War. The hat has a cavalry hat cord for enlisted men.

The left side has an ostrich plume attached (Fig. 91). This view also shows a more pronounced slope or angle to the rise of the crown front and back, than there is in Figure 90, side to side. This difference is possibly due to the definite oval shape of the body of the hat. The look is quite dramatic. The hat also has 1/2" wide hat band ribbon that is wrapped around twice to make it 1" wide.

The interior of the hat is lined on the sidewalls with off-white colored paper with a random line pattern. This gives it the appearance of faint marbleizing. The crown lining is gone and the linen over hat board is clearly visible (Fig. 92). The 1 7/8" wide sweatband is of cotton turned light brown with sweat and age.

There is an identification, hand written on a slip of paper glued in the crown (Fig. 93). It reads, "This hat was worn by Col. A. W. McDonald, 7th Va. Cavalry 1861." Col. McDonald was the officer who organized the 7th Virginia Cav. as the war began. He was old and continued as head of the regiment for only a brief period. One of his officers, Turner Ashby, became the Colonel, upon McDonalds resignation and retirement.

91

93

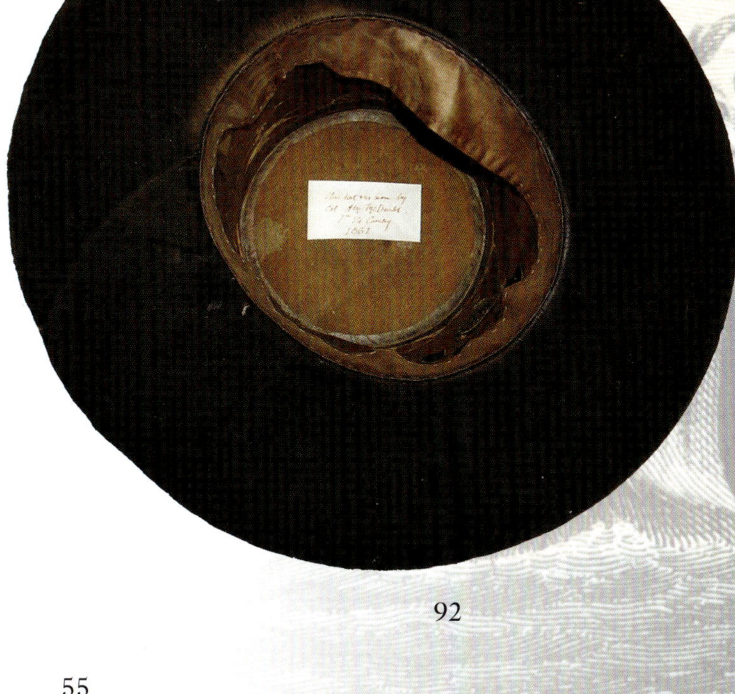

92

94

Confederate (Regulation) Officer's Kepi

This confederate kepi is one of the most desirable pieces of Civil War Headgear for any serious collector, and it is one of the truly outstanding pieces in the Budny Collection.

Made of blue gray wool and piped with gold lace, it is very striking. The dark blue, 1" wide band around the base is also wool (Fig. 94). The single strand bullion lace is in a quatrefoil pattern on the crown, it is along the upper edge of the dark blue band, it encircles the crown, and it is in vertical stripes on each side plus in front and in back.

The visor is bound, black leather, while the chinstrap is 5/16" wide bullion lace attached with small ball buttons (Fig. 95).

95

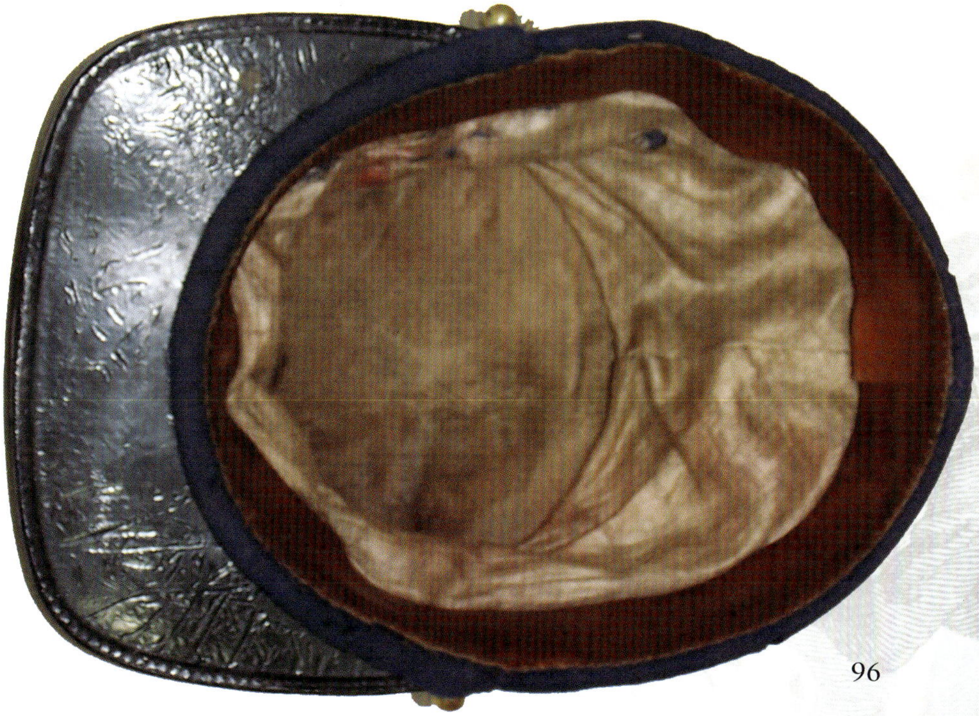

96

The cap is lined with light color polished cotton (Fig. 96). The 1 1/2" wide sweatband is of thin brown leather. Written on the sweatband in capital letters are the initials "P. J. G.".

97

98

Confederate Artillery Officer's Kepi

This bright red officer's kepi has all the right C.S. characteristics and features. The body is of red wool. It is trimmed with a single gold bullion lace. There is a bullion stripe all the way around the kepi 1 1/8" up from the base. Vertical lines of bullion extend up from this base in the front (Fig. 98), the back, and on each side (Fig. 100), to a circle of bullion around the edge of the crown. The crown is decorated with bullion quatrefoil as well (Fig. 99).

100

The visor is of bound black leather. The chinstrap is 1/4" wide gold bullion lace. The side buttons are small ball buttons.

On the inside (Fig. 101), the lining is complete with cream-colored linen in the crown and the same color cotton around the body. The complete sweatband is black, very thin leather.

99

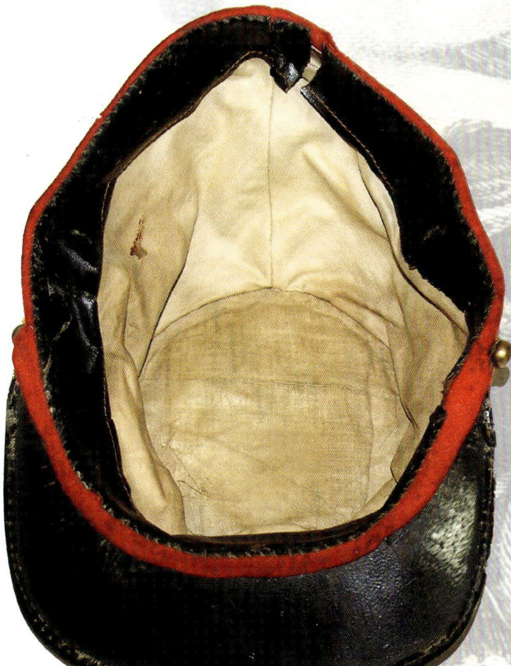

101

102

Confederate Enlisted Man's Kepi

This dark butternut colored kepi can be picked out of a crowd as confederate. Research leads us to believe that it is from a Maryland Regiment.

It is very plain as Figure 102 indicates. The body is a dark butternut wool, with a 1" wide dark blue band around the bottom edge. This is all hand sewn.

103

The side buttons are domed, floral brass two-piece buttons. The chinstrap is tarred leather. The visor is tarred and bound heavy cardboard (Fig. 103).

The lining of the crown is black oilcloth. The lining around the sides is near black in color cotton (Fig. 104). The sweatband is complete and made of black oilcloth.

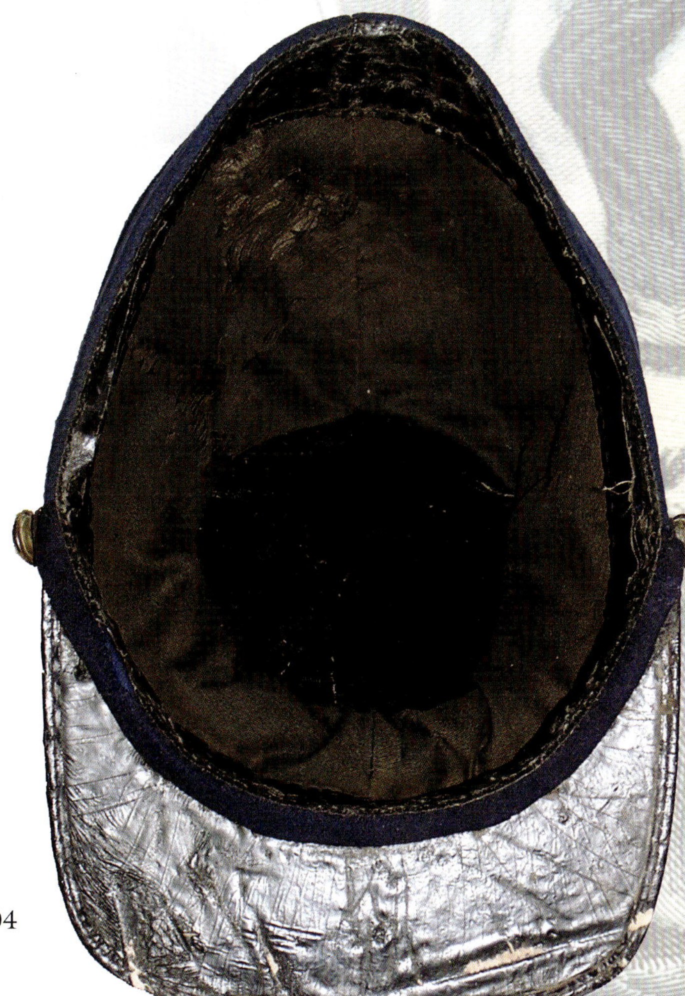

104

105

Civil War Identified Officer's Frock Coat and Slouch Hat

This frock and slouch are identified to 2nd Lt. Henry S. Foster, 6th NY Heavy Artillery. The 6th fought as part of the "Red Legged" infantry from the Wilderness to Appomattox Courthouse. Lt. Foster was wounded at Cedar Creek in 1864 (Fig. 105).

His patriotic stencil is of an American spread winged eagle. The banner in the eagle's beak has his name, "J. S. Foster" (Fig. 106).

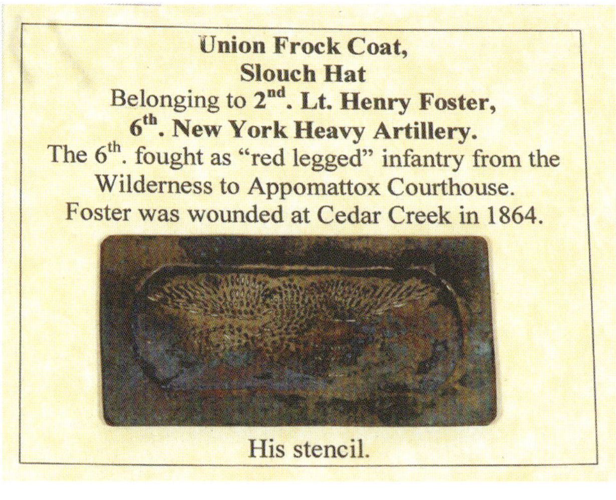

Union Frock Coat,
Slouch Hat
Belonging to **2nd. Lt. Henry Foster,**
6th. New York Heavy Artillery.
The 6th. fought as "red legged" infantry from the Wilderness to Appomattox Courthouse.
Foster was wounded at Cedar Creek in 1864.

His stencil.

106

107

108

The frock coat is a typical foot officer's frock of the Civil War, being made of dark blue wool. It has 2nd Lt. of Artillery shoulder straps (Fig. 107). It has nine New York state buttons down the front. The stand up collar is 1 7/8" high and lined with dark blue velvet.

The sleeves balloon out at the elbows and narrow down to 4 1/2" at the cuff opening (Fig. 108). Each cuff has only one of the three buttons it should have. The back coattail has all four of its buttons.

The coat lining is complete and has faded to a dark green (Fig. 109). There is an inside pocket on each side at the chest. These pockets are of brown cotton. The sleeve lining is of off-white cotton and "H. Foster" is printed in the left sleeve.

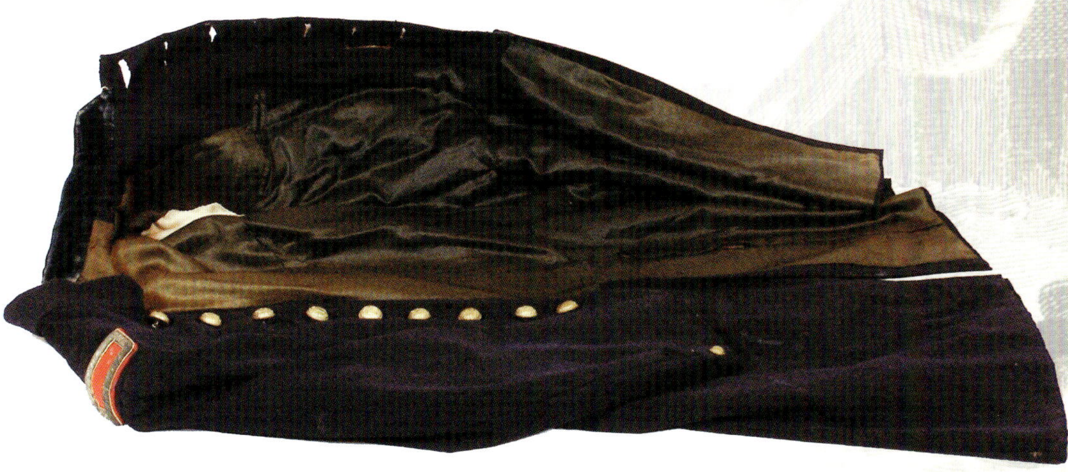

109

110

Lt. Foster's slouch hat has an artillery insignia on the front (Fig. 110). It is of false embroidered crossed cannons in stamped brass on a black velvet oval. The velvet has dye faded to dark brown. The brim is bound and here again, the black binding has mostly faded to dark brown.

The hat has a very wide (1") hatband of black ribbon (Fig. 111). The hat cord was originally black and bullion, but the black has faded to brown. The webbing remains black on the acorns.

The sweatband is 2 3/4" wide and made of a very thin black leather. Any lining the hat had, has long ago gone to pieces, and has disappeared (Fig. 112).

111

112

113

114

New York Civil War Great Coat

This great coat is made of heavy medium blue wool. It is double breasted with 14 New York state buttons on the front (Fig. 113). It is full length. There is no provision for attaching a cape and it never had one. Each sleeve has three NY buttons at the cuff (Fig. 114). Figure 115 gives a good look at the inside. It has a bright red wool lining extending through the whole body and tails. The sleeves are lined with off-white cotton fabric.

The coat has no stampings or marks. There are no pockets inside or out.

115

117

116

Civil War Drum and Drum Sticks

Mr. Budny purchased this drum and sticks to display with his drummer's uniform as seen in Figure 117. The drum has a Civil War label, which reads "John F. Stratton, Military Drums, 105 E. 22nd Street, NY." Stratton was at this address 1861-1864.

The decoration on the drum consists of brass tacks around, and vertically in line with the peephole, and a patriotic (red, white, and blue) painted pinwheel (Fig. 116).

After the Civil War, this drum was recycled and was used later, probably by a state militia unit. At that time (about the Spanish American War) a decal of an eagle was put on the drum. At first glance, all collectors dismissed the drum as of Spanish American War vintage, and gave it no consideration as being Civil War.

Knowing that it was Civil War because of the undeniable label, it was decided to remove the eagle and cover the spot where it had been with the patriotic pinwheel. The rationale was that: (1.) The drum was probably unpainted originally; (2.) The decal ruined its credibility; (3.) Removing the decal left an area where the original varnish and stain was gone; and (4.) It was better to cover the area with a generic folk/patriotic design than to leave it disfigured.

The drum measures 11 3/4" high and 14 5/8" in diameter. The low height plus the peephole being off center indicates that the drum was cut down slightly from its original height, probably at the same time the decal was added.

118

Civil War Federal Leather Shako – Imported from France

At the beginning of the Civil War these shakos were purchased in France, shipped to the US, and are known to have been issued to 5 northern regiments. These were the 49th NY, the 72nd NY, the 62nd PA, the 83rd PA, and the 18th Mass. They are made entirely of leather, finished in black, and were very hot and heavy to wear.

The front plate (Fig. 118) was produced for the American market and is quite attractive, featuring a large American eagle and an infantry horn. The pompons that were shipped with the shakos were green, which did not fit with the American scheme of things. These are seen with a variety of colored pompons and it's difficult to say that any are original to the hat they are on.

Each side of the shako has a metal vent near the crown and each side is reinforced with V shaped bands (Fig. 119).

The underside of the visor is green (Fig. 120). The 3" wide sweatband is complete as are the chinstraps, which are pictured coiled up inside the crown.

119

120

121

Civil War (Non-Regulation)
Federal Artillery Officer's Slouch Hat

This slouch hat is similar to and typical of many seen in Civil War photos. It is made of heavy pressed black wool felt and is quite stiff and rigid.

On the front, it has false embroidered, stamped brass crossed-cannon insignia (Fig. 121). This is tarnished green with age. The gold bullion officer's cord has very large acorns and it is dark with tarnish and years of exposure.

The brim would be 4" wide, except that it is rolled up all around to 3" and bound. The hatband is 1 1/8" wide. The binding on the edge and the band are the same material which was originally black. Now they are dye faded to a dark burgundy (Fig. 122). The color seems almost appropriate for an artillery hat.

There is no lining or sweatband in the hat (Fig. 123). The shadow from the sweatband indicates that it was 1" wide.

122

123

124

Civil War Tri-Corner Hat

This piece of militia headgear is from the Civil War and the state of Massachusetts. There were local New England units at the outbreak of the Civil War that wore Revolutionary war styled uniforms. Some of these enrolled in regiments from their home state at the outbreak of the war and the first call for volunteers. One such unit was from New York, one was from Connecticut, and one was from Massachusetts. There may have been others. This hat is from the Massachusetts unit.

The hat is the classic tri-corner shape and style (Fig. 125). It is 4" high and 14" on each side. It's made of good quality beaver felt.

On the left side is a large black ribbon cockade with a double stripe of 1/2" bullion lace and a large Massachusetts Civil War button (Fig. 124). A bright red feather plume extends flat over the crown.

On the inside, the hat has a complete black silk lining and a 1 5/8" wide burgundy leather sweatband (Fig. 126). There's a very small piece of paper with the date "1861" glued on one side of the sweatband.

125

126

127

US Civil War Rear Admiral's Visor Cap

In 1863 the US Navy changed a number of uniform regulations from what they were since 1861. The 1861 regulations called for the insignia on Rear Admiral's caps to be a single star in a wreath. The 1863 regulations changed this to two stars in a wreath. This hat may well be the rarest piece in the Budny collection because even the Naval Academy Museum doesn't have an example of this cap.

Provenance: This visor cap was formerly in the Paul DeHaan Collection. The DeHaan Collection has for many years been recognized as one of the premier Civil War Naval Collections in the nation. On at least two occasions, this hat was featured in award-winning displays presented by Mr. DeHaan at Major shows.

Figure 127 gives us a good look at the new, in1863, regulation insignia. This example has two brass stamped and gold washed stars on a felt oval with an embroidered wreath. The cap is made of dark blue wool.

Naval officers caps are very distinctive in profile (Fig. 128) that makes it easily distinguishable from all other American military hats with visors. This has US Naval officer's side buttons of the Civil War.
The chinstrap is black leather. The visor is also black leather with a bound edge.

The crown of the cap is padded and soft and not hard and stiff, like a kepi. On the inside (Fig. 129), we find that the crown lining is quilted. The lining is complete and is of dark brown cotton. The crown also has the remains of a British label, which is too faded and worn to be read. Estimates are that half of the US naval uniform items and insignia were imported from England during the Cival War. The sweatband is complete, made of brown leather and is 1 3/8" wide.

128

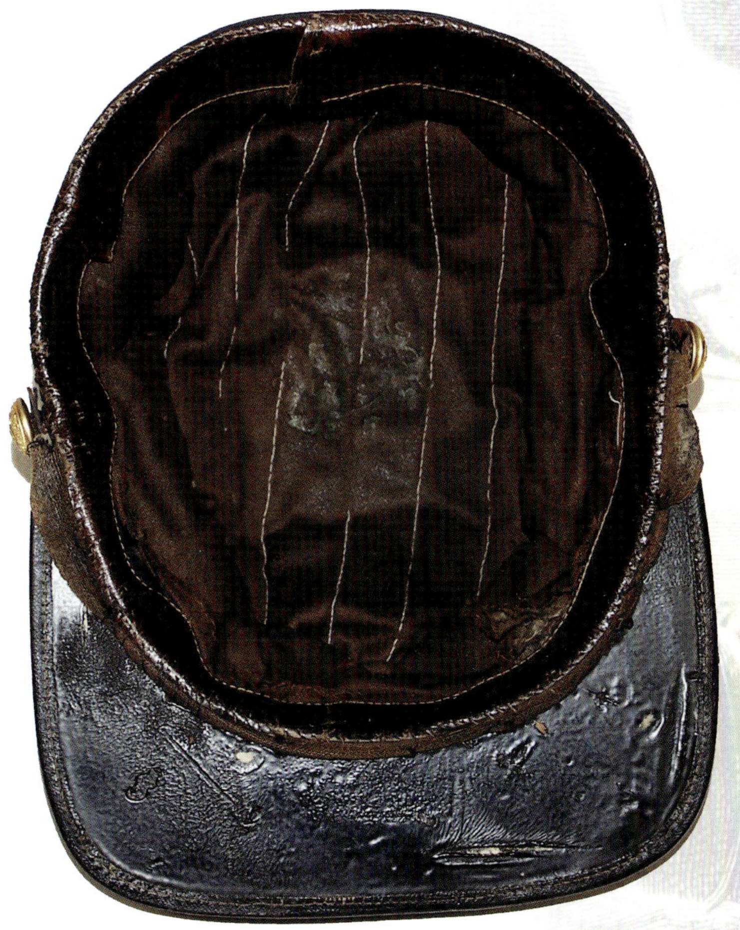

129

130

Civil War (Non-Regulation) Federal Artilleryman's Slouch Hat

This is a good example of the rounded top "plug" hat worn by many soldiers in the Civil War. The brim is not bound. It's just a very cheap, but serviceable and very comfortable hat of the kind often worn by enlisted men.

The hat is made of very soft and pliable black wool felt. It is so soft, it's difficult to make it keep its shape. It can be readily formed with the crown creased, bowed up (Fig. 130), or pushed down to appear almost flat. It has large, regulation stamped brass crossed cannons for insignia. The red cord is typical for enlisted slouch hats of all styles. The brim is 2 7/8" wide and curls up at the edge. There is a narrow (1/4") black hatband under the cord (Fig. 131).

Being cheaply made, this never had a lining. The brown leather sweatband is complete and 1 3/8" wide.

131

132

133

G.A.R. Straw Hat

G.A.R. hats are of many types. This summer straw hat is typical of the vet's interest in comfort, more than formality.

This was worn by a veteran artilleryman (Fig. 133), and likely he was enlisted, not an officer, because of the red wool cord. The front has a standard G.A.R. in silver finish inside a brass wreath insignia.

The hat has a 5/8" wide black hatband (Fig. 134). The brim is 3 3/8" wide.

The most unusual thing about this hat is on the inside. Figure 135 reveals a sweatband made of a burgundy colored ribbed cloth 1 1/8" wide. The same material is sewn around the connecting seam between the crown and the sidewalls, apparently to strengthen this area. This inside feature was unknown to the authors prior to this example.

134

135

136

Civil War Forage Cap Crate

There were three of these crates discovered in an Albany, NY warehouse in about 1990. This one was acquired by Robert Walter and became an important piece in his well-known collection. When the Horse Soldier in Gettysburg sold the collection, the crate became part of the Kenneth Budny Collection. It measures 21" high, 35 3/4" wide, and 20 1/2" deep (front to back). It is complete with its original lid and all the paint is original.

Figure 136 shows the front side of the crate with the lid on. Stenciled in white is "J. G. Cotrell & Son – 46 State St. – Albany, NY." Cotrell was at this address from 1858-1864. The opposite side is marked the same way. Seen with the lid off in Figure 138, it is divided in half by a partition.
The lid is standing on end against the right end. The entire crate is made of pine.

Each end of the crate is marked "200 U.S. Caps – 46 State St. Albany, N.Y."

137

138

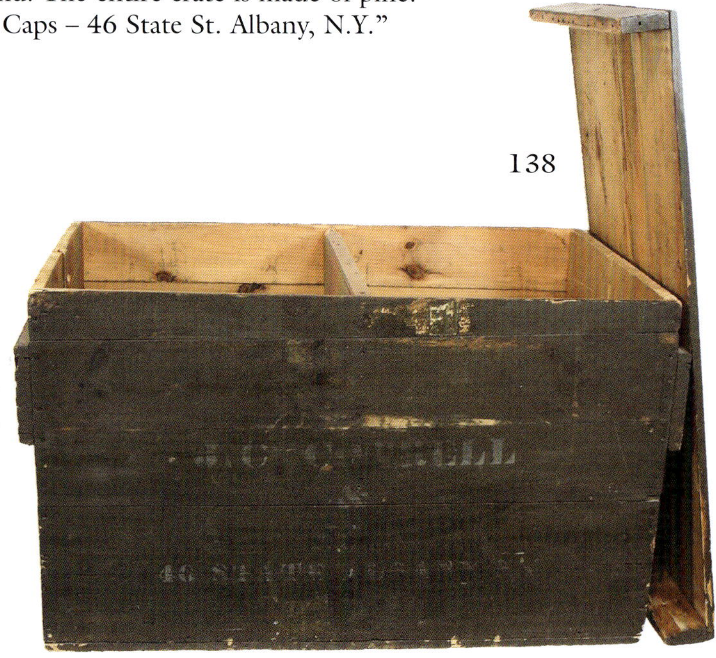

139

140

Civil War, Federal Officer's Waist Belt – Red Leather for Artillery Officers

This red belt is a seldom seen variation of the standard black leather officer's belt. The plate is full length and complete (Fig. 139).

A close-up of the plate reveals that it is of high quality and retains over 50% of its original gilding (Fig. 140).

141

Ancient and Honorable Artillery Company Waist Belt – circa 1870-1885

This belt is red leather. The plate and adjustment buckle are silver plated over brass. In the center of the plate is the seal of the company (Fig. 141). The Ancient and Honorable Artillery is the oldest continuous military company in the US. They were founded in 1638 in Boston, and continue to the present.

142

Civil War Period Hat Blocks

With his interest in headgear, Mr. Budny began to purchase hat makers forms from the Civil War Era. These hat blocks eventually came to number 7 and became a very unusual collection within a collection.

Figure 142 pictures three of the hat blocks. On the left is one with an unusual shape that is reminiscent of the form of a kepi from the side. It is unmarked and made of a very lightweight wood, perhaps balsa.

In the center is a low profile block that is again in the form of a kepi. The bottom of this is stamped in 3 places: "23 1/2", "Grand", and "F". It is made of pine.

To the right is a taller domed block made of pine. It is the kind used in making men's slouch hats. It is stamped on the bottom "881", and "21 1/2".

143

In Figure 143, we see on the left, a two-piece block that is very unusual in shape. It appears to be for a kind of shako or helmet. It is marked "22", "3121", "F", and "H. F. Block Co. N.Y.," and is made of pine.

At the center is a large domed block used to make men's slouch hats, this is marked on the bottom "22 1/2" and "798". It's very lightweight, and probably balsa wood.

To the right is a block in the form of a forage cap. It is unmarked and made of pine.

All of these hat blocks have many, many, working marks made by needles and pins.

The seventh hat block is the one for havelock's, dated "1864," and pictured earlier on page 41.

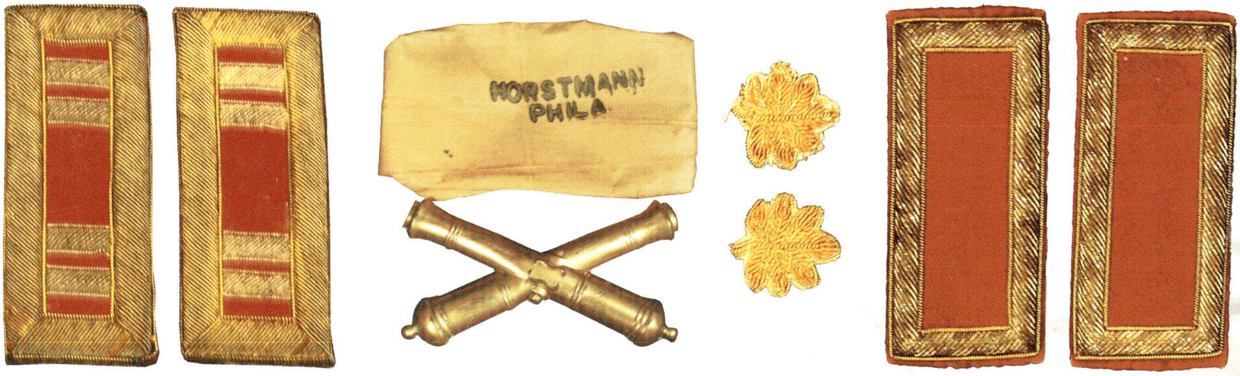

144

From left to right in Figure 144 the collection includes:
A pair of Civil War Captain of Artillery Shoulder Boards.
A pattern 1858 enlisted artillery hat insignia with its original "Horstmann, Phila" paper wrapper.
A pair of gold bullion oak leaves. (These are Civil War and suitable for attachment to shoulder boards of a major in any branch of service.)
A pair of Indian Wars shoulder boards for a 2nd Lt. of Artillery.

145

In figure 145 from top to bottom on the left we have:
A pattern 1858 enlisted artillery hat insignia made of stamped brass (3 1/4" wide).
A pattern 1855 militia enlisted artillery hat insignia made of stamped brass (2" wide).
A pattern 1858 militia enlisted artillery cap insignia made of stamped brass (1 1/2" wide).
In figure 145, from top to bottom on the right are:
A Civil War officer's embroidered artillery insignia on black velvet. The square of velvet would be cut in an oval or in an outline around the crossed cannons before being sewn onto a kepi or forage cap.
An embroidered artillery officer's insignia for a slouch hat. A metal unit number would be attached to the center red circle. This is Civil War, and measures 3 1/4" x 2 3/8".

146

147

Civil War Shoulder Scales

These were made to be worn on shell jacket shoulders for Artillery or Cavalry (Fig. 146).

Mexican War – Silver Epaulets

This pair, for a Lt. Colonel of Infantry, have silver eagle "I" buttons and silver oak leaves. One has its original ribbon tie for attaching. The undersides were white silk and most of the silk is missing (Fig. 147).

148

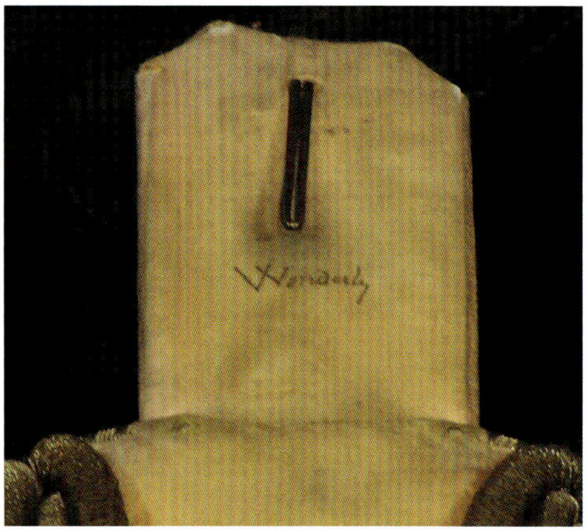

149

Identified Civil War Epaulets

This pair of epaulets is marked on the underside "WONDERLY" in ink. He is Captain David Wonderly, 32nd Pa. Vols. These were purchased out of his estate in Grand Rapids, Michigan. He moved there with his wife and child after the Civil War.

Figure 148 shows the pair with very heavy fringe and no buttons. They do not appear to have ever had buttons. They attach by means of the long wire hook seen in Figure 149 and so a button would be only decorative and difficult to attach.

Civil War Epaulets

This pair is in their original kidney shaped tin. Most of the finish is worn off the tin, but the inside has a full quilted red silk lining. These are for a Lt. Colonel of the 1st Artillery. They are complete.

150

151

Civil War Artillery Epaulets

These have a "Horstmann Bros. & Allien, New York" label. They are for a Major of the 5th Artillery. They are complete and in their original tin, which has 90% original japanning (Fig. 151).

Civil War Epaulets

For a Lt. Colonel of the 1st Artillery, these are complete and unmarked. The underside is yellow silk. They are in their original japanned tin (Fig. 152).

152

153

G.A.R. Grouping from the 9th New York Heavy Artillery

This is an assembled grouping. Because he had the identified ninth NY Heavy Artillery forage cap and tintype, Mr. Budny began to collect items related to the 9th. This grouping is a result of that interest.

The grouping includes:

1. A cdv of Col. Joseph Welling, 9th NY Heavy Artillery.
2. A ribbon from the 31st Annual Reunion of the 9th NY H.A. at Walcott, NY in 1905.
3. A large Reunion Ribbon for the 1897 gathering in Buffalo, NY by the 9th NY H.A.
4. A 1909 ribbon for the 35th Annual Reunion of the 9th NY H.A. at Auburn, NY.
5. A 33rd Annual Reunion Ribbon held at Marion, NY in 1907 for the 9th NY H.A.
6. A ladder badge of the 9th New York Heavy Artillery with a 15th corps badge at the bottom.

154

Civil War and G.A.R. Grouping of Artillery Items of Isaac Wyant, Battery K, 5th US Artillery, Artillery Reserve Army AP.

The grouping includes:

1. A letter from Wyant dated Mar. 15, 1863 with it's envelope.
2. A 1/6 plate Ambrotype of him in uniform with 2 civilian friends (all with ID on the back).
3. His Civil War artillery hat cord for enlisted men.
4. His G.A.R. medal.
5. A 48 star G.A.R. flag. The dealer Mr. Budny purchased the group from, added this to the grouping.

155

Two Kepi Rain Covers

Mr. Budny had in his collection, two kepi rain covers. One is pictured in Figure 155 on the far left displayed on a kepi hat block. The other one is pictured to the far right. One is slightly larger than the other, but beyond that they are very alike. Made of black oilcloth, they might fit almost any kepi from the late Civil War through the Indian Wars. Some were made for use on fraternal kepis of the same time span. We are unaware of any martially marked ones, and so there is no way to know for sure or guarantee military usage.

Child's Boots - circa 1860-1865.

These boots, pictured in Figure 155, are small (8" long, toe to heel, and 11 1/2" high from heel to top of the shield cut front). Made of black leather, these saw much use; probably passed down from one son to another. They have been re-heeled and half-soled long ago. Originally, they were wood pegged with square toes, as they should be for the period. The half soles are wood pegged, while the toe ends and heels are done with nails. Mr. Budny displayed these boots with his Civil War Drummers Coat and Hat.

156

1889 G.A.R. Flag

This 42 star flag is framed (Fig. 156) and has printed on two of the white stripes, "G.A.R." and "1889". The frame measures 24 3/4" x 17 5/8". It is a printed flag on thin linen and was made for an 1889 reunion.

Paul Millikan

Paul Millikan brings 42 years of experience, acquired knowledge, and expertise to this publication. He is by profession a historian, having both undergraduate and graduate degrees in American History, he taught American History including special period courses, such as Civil War and Reconstruction, at the college level for 32 years. Before that he taught 6 years at the high school level.

Professor Millikan also served for six years as director/curator of George Washington's Boyhood Home at Fredericksburg, VA.

In addition to his academic credentials, he has collected, studied, researched, written, and dealt in Civil War memorabilia, for those 42 years. One of his special interests has been Civil War and earlier headgear. Over the years he has owned, studied, or handled literally hundreds of pieces of headgear.

Since 1982, he has been author and publisher of The Manual of Arms annual catalog of weapons, accouterments, and Americana. He has a national reputation through his outstanding catalog and regular participation in Civil War shows. In addition he has, for 17 consecutive years, been a participant in the prestigious Colonial Williamsburg Antiques Forum.

Since his retirement from college teaching in 2000, he has been involved in publishing, writing, research, living history video production, and public speaking.

He is a life member of the Michigan Antique Arms Collector, a member of the Organization of American Historians, the author of several articles and publications, as well as a regular participant in several Civil War shows around the country.

Richard Listenberger

Richard Listenberger, graphic designer and visual communications consultant, has a passion for military history that spans 30 years as a collector, researcher, writer, dealer, and specialist in the restoration of fine art and militaria.

With a BFA degree in advertising and marketing from John Herron Art Institute, Indiana University, Richard has owned and managed his own business since 1975. As well as major corporate clients, he has provided award-winning creative services for a wide variety of history and Civil War publishers, museums and military organizations. He has created and merchandized his own Civil War products and publications in addition to writing and producing a documentary play on the American Revolution.

Richard's specialty is researching and collecting American military headgear, uniforms, art, swords and accoutrements of the Revolutionary War through 19th Century, with specific attention to Federal period military fashion of militia and volunteer regiments.

He is a member and past President of The Antique Arms Collectors Association of Indiana; member and past Commander of The Benjamin Harrison Camp of Sons of Union Veterans; and long-standing member of The Ohio Valley Military Society.

THE MANUAL OF ARMS PUBLISHING

For further information concerning forthcoming books in this series, visit our website:
www.themanualofarms.com